Ŋilimurruŋgu Wäyin Malanynha

SIENA STUBBS

Pronunciation Guide

We speak Yolŋu Matha, the language of the Yolŋu people. There are six letters in Yolŋu Matha that don't exist in English, this is because some sounds in Yolŋu Matha are unlike any in the English language. As it has become a written language new letters have been included to symbolise these sounds:

ŋ pronounced 'ng' as in sung

ä pronounced like the 'a' in father

d̲, t̲, l̲ and n̲ retroflex consonants formed with the bottom of the tip of the tongue curled up to the roof of the mouth

' the apostrophe represents a sudden cutting off of the previous letter.

rr (double r) is always rolled (vibrating tongue)

In pronouncing words in Yolŋu Matha the emphasis is always on the first syllable.

Listen to Siena pronouncing the Yolŋu Matha bird names on the Magabala Books website:
https://www.magabala.com/contemporary-non-fiction/our-birds.html

Opposite: D̲amala, WHITE-BREASTED SEA EAGLE, Dhuwa

This book is dedicated to my Aunty Sash, my
Evil Queen, who gave me my first proper camera.

My name is Siena Mayutu Wurmarri Stubbs and I am a fifteen-year-old Yolŋu girl who loves photography. My malk, or skin, is Gutjan, I am a Gumatj girl from the Yirritja moiety.

About three years ago, when I was twelve, I was determined to find the perfect camera for me. I was tired of zooming in with a blurry iPad and I was worried about losing memories captured in digital photos. So when my Dad went down to Sydney, he asked my Aunty Sash where he could buy a good quality camera. Coincidentally, she had one packed in her cupboard and she was more than happy to give it to me. In fact, she was so excited, when she tried to bring it down, it fell on her and gave her a black eye.

In the Yolŋu culture, everything – from a body of water to a specific type of tree – is divided into two moieties: Yirritja and Dhuwa. The birds give themselves their Yolŋu names by the sounds they make. There are so many birds in Arnhem Land it is difficult to keep count, but here are some I was able to capture.

Mayutu.
Siena

Mundjirr

BLUE-FACED HONEY EATER

Dhuwa

This was my first photo of a bird with my new camera. I was amazed when I saw the vibrant blue against the pink of the blossom. I remember taking this early one morning. My Dad and I were completing the circuit of our walk when we spotted the Mundjirr trying to find the nectar of the powder-puff flower as the sun was rising. Luckily, I had recently been given my camera and had it hanging around my neck. So I simply framed and captured this photo.

Murriyil'

TORRESIAN PIGEON

Yirritja

The Murriyil' comes from the Torres Strait Islands to Arnhem Land every wet season. If I stand on the beach of Ganirri'mirri and look across to the island Barrpira, I can see white dots in the trees. These are the pigeons resting during the night when visiting Arnhem Land. The Murriyil' makes a sound as though it is saying, 'Who who, who who'. We call them the 'who people'. When I was growing up we used to make knock knock jokes with them and wait for their response.Try this next time you see a Murriyil'.

Guluwukbuk

PHEASANT COUCAL

Dhuwa

Guluwukbuk is the perfect example of a bird that is named by its call. When Mum, the women of my family and I are out gathering borum or fruit, we can hear 'Guluwukbukbukbukbukbuk'. Sometimes when I wake up on a Saturday morning, if I look out from my back veranda, I can see the Guluwukbuk hopping around. This is because, although this beautiful bird has amazing long feathers, it isn't the best at flying and can fly only a couple of metres at a time. When I see the Guluwukbuk , it is either on the ground or struggling to fly from one branch to another.

Garrukal'

BLUE-WINGED KOOKABURRA

Dhuwa

Most people know of the Laughing Kookaburra. But the species that is found up here in North East Arnhem Land is the Blue-Winged Kookaburra. Instead of laughing they have a throaty chuckle with some screeches here and there. This can be noisy some mornings when the whole family is joining in on a chorus of 'Garrugarrukalgarrukalkalkal!' Kookaburras are a type of kingfisher, which explains the blue on the wings of the Garrukal'. I see them high on branches scanning the ground for lizards and snakes. Sometimes, they are high up on the electric powerlines.

Bilitjpilitj

RED-WINGED PARROT

Yirritja

When I observe the beautiful colours of Bilitjpilitj for a while it makes me appreciate how stunning nature is. While its body is lime green, it has scarlet red under its wings and a dash of deep blue on its bottom. The Bilitjpilitj flying pattern makes them easy to spot: they have a looping flight and widely spaced wingbeats. These characteristics make them fun to find. They are quite rare and shy but when the grass seeds are out, they gather in excited groups. Parrots can be found all around the world and it's nice to think we have our own attraction in our little corner.

Ganyiri

BRAHMINY KITE

Yirritja

Ganyiri is an amazing flyer. It soars high in the sky spotting its prey. I love it when Ganyiri soars and I can see the sun shine through its feathers. Mum and Dad have always been telling me about the life that surrounds us. I can still remember when Dad told me there was a bird called a Brahminy Kite. I thought it was cool that a bird was called a kite. Then I realised how much it actually acted like a kite. Ever since then, I've always admired Ganyiri. It was even my favourite bird for a very long time.

Ma<u>l</u>awi<u>d</u>iwi<u>d</u>i

BROWN GOSHAWK

Yirritja

I spotted this bird, alone, in my backyard. I have never seen it surrounded by any other birds. I sense it likes to be on its own. Ma<u>l</u>awi<u>d</u>iwi<u>d</u>i does look quite scary with its big talons, sharp eyes, curved beak and muscly legs – so it might be a bit intimidating for the smaller birds to be within its reach.

Ŋaṯili

BLACK-COCKATOO

Dhuwa

Every wet season, I know there is rain on the way when I hear the call of the Ŋaṯili. Sure enough, within a couple of minutes, a big grey rain cloud can be seen hovering nearby. The Ŋaṯili loves to eat the fruit of the Mäṯpana. The Mäṯpana is the Indian Almond tree that has large, tough fruit with tiny seeds inside, which taste like almonds. With its skilled beak and tongue, the Ŋaṯili manages to break through the hard shell and feast on the seeds. Gnawed shells are under every tree. I was told at a young age that if I look at this beautiful bird, there is a risk my mother could die. This must have been told to the younger generations to protect the Ŋaṯili.

Wirriwirri

RAINBOW BEE EATER

Dhuwa

The Rainbow Bee Eater is a beautiful bird. It hangs around in groups. Sometimes, up the street from my house, I can see a group resting on the road in the golden sunset. I can hear the hum of 'wirriwirriwirri' as the group chatters away. When a Wirriwirri catches an insect, it hits it against a tree to kill it then eats it in one bite. This bird's colour is mind-blowing. Its name really reflects its appearance – from afar I really can see the colours of the rainbow.

Djirrindidi

KING FISHER
Dhuwa

Even though you cannot see it in the photo, the main reason I love this bird is its gorgeous colour. This is my favourite colour range, like an ocean wave at its peak just before it crashes. The turquoise, blue and green make me feel as though I'm surrounded by colour, like being in the ocean. Djirrindidi also has great shape in the curve under its beak as well as the shape of its head. I think these amazing shapes help it dive into the water smoothly to catch ŋatha, food.

Ŋerrk

SULPHUR-CRESTED COCKATOO

Yirritja

This intelligent bird can be seen all around Australia and can be quite cheeky. I think I captured the perfect moment when taking this picture, with the iconic flash of yellow on its head. I see Ŋerrk everywhere I go. I can also hear them everywhere I go with their shrieks of liveliness. I generally hear them as I wake up. We used to feed them every morning with seeds that we placed on the rock just on the edge of our small cliff. A group of around thirty Ŋerrk would come and perch. It was quite a nice sight but we had to stop because too many snakes came to eat the mice who also fed on the seed.

Ṉama’

JABIRU

Dhuwa

The Jabiru is such a huge bird, it looks like a dinosaur. I’m sure it is somehow related to the Pterodactyl. It’s so big that it even looks huge from hundreds of metres away. It is rare to see Ṉama’ but when we go hunting, my family and I sometimes see them stalking the sand flats. The last time we saw Ṉama’ was when we were coming back from my homeland, Bawaka, otherwise known as Port Bradshaw. This homeland was passed down to my brothers, sisters and I from my Mum’s sister’s husband. It is now my brothers’ job to pass it down to their children.

Djet̠

OSPREY

Yirritja

I was amazed when I was able to capture this photo. This was taken at a moment's notice. I was not prepared. I just took the photo and wished for the best. The Maḏarrpa clan story we are told is that Djet̠ was a little Yolŋu boy who went hunting with his family, the same as we do. One day, he found a fish, cooked it up and ate it all by himself, even though his father wanted him to share it. With this, his father went hunting and caught all the delicacies that we hunt for. He returned with his catch, cooked it up and shared it with the family – except for Djet̠. When Djet̠ reached for a piece his father smacked his hand. This was his punishment. He then cried and cried until he grew feathers and flew off. Now Djet̠ catches its own fish and eats them raw. When we go hunting, we always offer a spare fish to the Djet̠ flying nearby.

Ḻindirritj

RAINBOW LORIKEET

Dhuwa

The Rainbow Lorikeet is a very colourful and chirpy bird. It chitter chatters in the morning and afternoon sun every day. I love its vivid colours, which almost pop off the page. This is another bird that I have adored for a long time. I think my younger self appreciated how a bird could be such art. It's as though someone painted it. This is another one of the birds that tries to wake me up – even on weekends when I try to sleep in a bit longer than dawn. Most days, Mum puts a squeeze of honey into our bird bath and that's all it takes to draw Ḻindirritj in for a drink, like my Momo and her friends do every Friday for lunch!

Djiḻawurr

ORANGE-FOOTED SCRUB FOWL

Yirritja

This bird is known for making very big nests, which are more like mounds. I always see them in pairs scratching around. It is a sacred spirit of the Gumatj, my clan, and the spirit of Bayini. It is said that the women of the Gumatj tribes sometimes transform into Djiḻawurr. Bayini is the spirit of Bawaka. She was a Makassan woman who came to Bawaka when Yolŋu traded with the Makassans. We traded trepang, knowledge of the sea and where to anchor, for pots, swords, axes and cloth. This relationship between the two cultures continued for centuries. I know that I have Makassan blood in me and it makes me excited. If you are walking through the jungly bush looking for borum or fruit in the Midawarr (harvest) season, you can hear Djiḻawurr talking to each other.

Muthali’

PACIFIC-BLACK DUCK

Yirritja

Surprisingly I often see Muthali’ happily floating on the sewerage ponds at Yirrkala. We pass these ponds when we drive down to Garriri, a beach nearby otherwise known as Rocky Bay. We usually come down for afternoon outings after school or for a walk. Other times, we go hunting for mud mussels out the back of the beach where the mud flats are, though we’ve got to be careful for the Gatapanga, Buffaloes, that like to hang around there too. Muthali’ are well known for their personalities in many cultures around the world but I know them for floating on the little ponds on the drive out to Rocky Bay.

Djulwaḏak

FIGBIRD

Dhuwa

I usually hear this bird before I see it. It has a kind of loud, high-pitched, deep-throated, tuneful chirp. It can make you jump! And then look up into the tree and see that beady eye surrounded by a bright red eye patch. As the English name suggests, the Figbird likes fruit. When I see one it might have a djin̲'pu, sandpaper fig, filling its mouth. I was able to capture this bird's love for singing. The way it tilts its head back when it sings is inspiring. This is how I would sing my favourite song.

Birrkpirrk

PLOVER

Dhuwa

This bird has a bad reputation. The poor thing makes its nest on the flat ground, which means it has to be very protective of its young. To walk to our shop I have to cross Rika Park and ever since I was a small girl Birrkpirrk has always made a big fuss every time I cross its territory. But I respect it because it is just being a good parent. It's funny, I used to think this bird constantly had its mouth open – but it is just the flap that hangs from its cheek!

Gurrumatji

MAGPIE GOOSE

Dhuwa

When we hear Gurrumatji honking overhead at night we know the season is on to hunt it. This bird is delicious, a feast for the families. The birds congregate on the golf course at Nhulunbuy, which is the nearby mining town. They love the green grass and seem to know they are safe there. My family gets frustrated every time we drive by because we can't hunt them on the golf course and have to go into the swamps and wetlands to catch them. The smell of the lily leaves and roots crushed by their feet as they greedily munch on water chestnuts is called läka and we cry for this smell.

Gurrutjurtjur

BROWN FALCON

Yirritja

This bird is Gumatj like me. We are the fire. Bäru was changed from a man into the Saltwater Crocodile by fire. Gumatj clan sing Bäru and fire and also Gurrutjurtjur, along with many other things. Whenever we clean the country with fire in the Rarrandharrar season (late-dry, October to December), these Brown Falcons join large flocks of Whistling Kites and Black Kites to catch the little animals and insects scurrying from the slow-moving line of fire. If a fire dies out, Gurrutjurtjur swoops down to pick up a smouldering stick and drops it somewhere else to restart the fire.

Gany'tjurr'

REEF HERON

Yirritja

For Yolŋu this bird is the Yirritja hunter stalking its prey in the shallows with his spear. He crouches down and creeps up low. When the fish is near he stands tall with his spear poised and then strikes! If you watch Gany'tjurr', he does the same. When the tiny waves wash in and the foam obscures the fish's view, he stalks in low. As the water stills and clears before going out, he picks his target and spears it with his beak. He is djambatj, a deadly hunter.

Acknowledgements

Thank you to the Evil Queen (my Aunty Sasha) for giving her Naughty Princess my camera and paying to have those first photos printed. To my Mum, Merrki Ganambarr-Stubbs, for giving me the knowledge of the identity of the birds and my Dad, Will Stubbs, for going on walks with me. To my faithful dog Binga Binga for making me smile. And my Ma'mo (grandfather) John Stubbs for passing on his love of the bush and my Momo Romey Stubbs who I love to the moon and back. And to my Manyi and Ngathi, Gaymala and Mowarra for loving me into this beautiful world. I thank the Charles Darwin University for their amazing online dictionary at http://yolngudictionary.cdu.edu.au/ which is worth reading just for its own sake.

First published 2018, reprinted 2018 x2, 2019 x2
by Magabala Books Aboriginal Corporation
1 Bagot Rd, Broome (PO Box 668 Broome 6725)
Western Australia
www.magabala.com e: sales@magabala.com

Magabala Books receives financial assistance from the Commonwealth Government through the Australia Council, its arts advisory body. The State of Western Australia has made an investment in this project through the Department of Local Government, Sport and Cultural Industries in Association with LotteryWest. Magabala Books would like to acknowledge the support of the Shire of Broome, Western Australia.

Department of Local Government, Sport and Cultural Industries

Designed by Big Cat Design
Printed in China by Toppan Leefung Printing Ltd

Cataloguing-in-publication data available from the National Library of Australia

ISBN 978-1-925360-98-1